Kira
and
Lulu
visit the
Sun

BY
Kat Davidson

Kira and Lulu visit the Sun

By Kat Davidson

Illustrations by Kat and Karen Davidson
Graphic Assistance by David Senkiw
Font Copyright - Quicksand (Bold)

A project of Space Weather News LLC

www.SpaceWeatherNews.com

www.ObservatoryProject.com

ISBN: 978-0-692-06956-1

This book belongs to:

To those that find
amazement and wonder in
all that lies beyond
the skies.

It's a beautiful day – the Sun
is shining so bright!

Kira and Lulu are outside having a picnic;
cupcakes are such a delight!

"I have an idea that's never been done." She smiles and says, "Let's visit the Sun!"

Mom and Dad and little Brother
wave goodbye, as Kira and
Lulu head to their rocket to fly.

"Travel safe and be careful, there is a lot to learn.

Dinner will be ready when you return!"

As they flew into space the solar
wind blew,

with the earth behind
them; it was a beautiful view.

"We're here!" They quickly climb out of their rocket.

Its so bright!

"Grab your special solar sunglasses from your pocket!"

Kira said "Lulu, look over there!"

Time to pass Mercury, Venus,
and now, the moon!

It's the flare we saw leaving
the sun - when it gets to earth it
is very fun!

The energy from the sun creates
a colorful show, and the sky is filled
with a wonderful glow.

The solar flare has just arrived...
its time for the colors to dance in the sky.

It's been a long day,

time to turn out the lights.

Sweet dreams, Kira and Lulu,
time to say goodnight!